# Look at the Mechanism behind the Postulate of the Equivalence Principle

The Mechanism behind This Extremely
Important Postulate Has Been Revealed

ALSO BY BINGCHENG ZHAO

(Popular Science & Science)

*Why It's Difficult to Understand "A Brief History of Time"*

*Terminate the Controversy over the Big Bang Theory by Inspecting All Its Three Pillars*

*Dark Matter Is No Longer in the Darkness! (The Constituents of Dark Matter Have Been Revealed)*

*From Postulate-Based Modern Physics to Mechanism-Revealed Physics*

(Science & Religion)

*The Amazing Wisdom and Truth: GOD Can Change the Scales of Space and Time in the Universe*

*The Naked Truth: The Big Bang Theory Cannot Deny GOD!*

*The Hard Evidence: The Big Bang Theory Has No Way to Deny GOD! (After Inspecting All the Three Pillars of the Big Bang Theory)*

# Look at the Mechanism behind the Postulate of the Equivalence Principle

## The Mechanism behind This Extremely Important Postulate Has Been Revealed

Bingcheng Zhao

LOOK AT THE MECHANISM BEHIND THE
POSTULATE OF THE EQUIVALENCE PRINCIPLE

Published in the United States of America through Amazon's KDP

The copyright of this book is owned by the author of this book

Copyright © 2019 by Bingcheng Zhao—the author of this book

All rights reserved, including the right of reproduction in whole or in part in any form or by any means. No part of this book may be reproduced or transmitted in any form or by any means—electronic or mechanical, photocopying, recording, without the permission from the author of this book. No part of this book may be translated into any other languages without the permission from the author of this book.

ISBN-13: 978-1794187207

# CONTENTS

| | |
|---|---|
| FOREWORD | vii |
| Chapter 1: The Questions Unavoidably Pointing to the Law of Mass Doing Work | 1 |
| Chapter 2: The Law of Mass Doing Work—about Why Mass Has Energy | 5 |
| Chapter 3: Revealing the Mechanism of/behind the Mass-Energy Equivalence Equation | 9 |
| Chapter 4: The Mass Consumption | 15 |
| Chapter 5: Revealing the Mechanism behind the Postulate of the Equivalence Principle | 19 |

# FOREWORD

The theory of general relativity was very happy and excited after having been aware that the mechanism behind its famous postulate of the equivalence principle has been revealed. In fact, general relativity was so interested in this mechanism that it couldn't help continuing its discussion about this mechanism with the physicist mentioned in the book description of this book, whose name is Michael W. Moore, a reviewer of this book.

"Hi, Mike, would you please concisely tell me how the mechanism behind the postulate of the equivalence principle has been revealed? If so, I can read this book more effectively. To tell you the truth, for quite a long time, I have been constantly and persistently searching for the answers to the questions: is there the mechanism behind the postulate of the equivalence principle? And what is this mechanism if it really exists? But all my efforts have been fruitless until now," general relativity asked and explained.

"The main route along which the mechanism behind the postulate of the equivalence principle has been revealed is: the law of mass doing work → the mass consumption (a direct product of the law of mass doing work) → the mechanism behind the postulate of the equivalence principle," the physicist answered concisely.

"Aha! What? The law of mass doing work! I have never thought of this law. No wonder my methods and efforts didn't work," general relativity was very surprised, also somewhat regretful.

"Yes, the law of mass doing work is indeed the key to revealing the mechanism behind the postulate of the equivalence principle, though this revealing isn't the main purpose of this law," the physicist assured general relativity calmly.

"What's the law of mass doing work? What's this law about?" general relativity eagerly wanted to know this law.

"About these two questions, it's better for you to ask or discuss with the law of mass doing work," the physicist answered and suggested.

# FOREWORD

"That's a good idea. I'm going to have a talk with the law of mass doing work," general relativity gladly accepted the suggestion of the physicist. ...

"Hi, dear the law of mass doing work, I'm very glad to see you. Great thanks for your warm greeting," general relativity was very excited when it met this law.
"You are very welcome. I'm looking forward to seeing you; and it's my pleasure and honor to meet you," this law greeted general relativity keenly and enthusiastically.
"After having a very meaningful talk with a physicist, I eagerly want to talk with you. I have six questions to discuss with you, may I?" general relativity directly went to the topic and purpose.
"Surely, of course; I'm very delighted to discuss any questions with you. Moreover, I believe that I could learn many important things from you through our discussions, because you are a very great theory in science," the law of mass doing work responded gladly and warmly.

"My first question is: what's the most important concept of the law of mass doing work?" general relativity asked very curiously.
"It's <u>mass has the capability of doing work</u>," this law answered concisely.
"Ah! What? Mass has the capability of doing work! How could one think of such a concept?" general relativity was obviously surprised, even astonished.
"Please don't be so surprised. The important reason that one could think of this concept was inspired by several fundamentally important questions from the famous and great $E = mc^2$," this law responded calmly.
"Oh! Really?" but general relativity became even more surprised.
"Yes! Really, and surely! You'll see that in chapter one," this law answered calmly and assuredly.
"That must be very, very interesting! I'll read that chapter as soon as I can," general relativity reacted with great excitement and curiosity.

"My second question is: what's the most important question the law of mass doing work has answered? And how?" general relativity continued.
"It has answered the question of *why* mass has energy by revealing and showing that <u>mass has the capability of doing work</u>, as introduced and emphasized in the first three chapters of this book," this law responded directly.
"Great! This is remarkably important, and fundamentally important, also highly interesting, because the question of *why* mass has energy is definitely

# FOREWORD

and obviously a fundamentally important question in physics or in science, considering the great importance of the widely recognized fact that <u>mass has energy</u>," general relativity commented excitedly and approvingly.

"You are really insightful and highly visionary," this law praised general relativity genuinely.

"My third question is: since the law of mass doing work has answered the question of *why* mass has energy; since the mass-energy equivalence equation ($E = mc^2$) tells us mass has energy, could this law reveal the mechanism of $E = mc^2$?" general relativity asked with great curiosity.

"Yes, it could; because it has revealed this mechanism," this law replied concisely.

"What is it then?" general relativity asked hurriedly.

"As introduced in chapter three, this mechanism is, and turns out to be, that the $mc^2$ (in $E = mc^2$) is, and is equal to, the maximum capability of m doing work. That is to say, this mechanism shows that the $mc^2$ (in $E = mc^2$) is not only the total energy contained in a rest mass m, but also turns out to be the maximum capability of the rest mass m doing work. Moreover, this mechanism further shows that the very reason *why* the total energy contained in a rest mass m is equal to $mc^2$ is because the maximum capability of the rest mass m doing work is equal to $mc^2$," this law answered and explained.

"Ha! What? The $mc^2$ is also equal to the maximum capability of m doing work! This is amazing, really amazing, and wonderfully amazing! This really makes the $mc^2$ much more meaningful! This also greatly expands my field of vision," general relativity became very surprised and excited.

"My fourth question is: what's the core principle of the law of mass doing work?" general relativity asked further.

"The amount of energy in the mass of an object is measured and determined by the amount of work done by the object's mass," this law replied concisely.

"Such a core principle is rather rational thus pretty easy to understand, because it is quite comparable or very similar to that the energy of a body is measured and determined by the body's capability of doing work in classical physics," general relativity commented in an analogous connection.

"You're definitely correct. That's why the law of mass doing work is easy to understand, though it is quite new in concept," this law agreed and explained.

# FOREWORD

"My fifth question is: what's the core point of the law of mass doing work?" general relativity wanted to know more about this law.

"Its core point is: when a mass does positive work, the mass thus decreases; but when a mass does negative work, the mass thus increases," this law answered directly.

"Such a core point is also easy to understand, because there is such a simple comparison in classical physics: when a body does positive work, the energy of the body decreases; but when a body does negative work, the energy of the body increases," general relativity commented in an analogous mode of thinking.

"Your comment is definitely true. And your comment could make one realize that he or she could easily understand the law of mass doing work, though it is radically new in concept," this law replied approvingly.

"My last question, also the most important question to me, is: what is the mechanism behind the postulate of the equivalence principle?" general relativity asked with great concern and curiosity.

"This mechanism is the mass consumption," this law answered concisely.

"What's the mass consumption then?" general relativity asked more curiously.

"The mass consumption is the decrease in the mass of an object, because the object's mass is consumed (becoming less) due to its doing positive work when the velocity of the object is increased," this law replied.

"Aha, I see. The mass consumption is a product of the law of mass doing work. And the mass consumption always exists, works and manifests in the situation or condition that the postulate of the equivalence principle describes or defines; this postulate says that gravitational force has the same effect in increasing the velocity of an object as other traditional forces. No wonder the mass consumption is the mechanism behind this postulate," general relativity got and understood the answer it wanted most.

Bingcheng Zhao

# Chapter 1

# The Questions Unavoidably Pointing to the Law of Mass Doing Work

## [The Window on This Chapter]

"Since the law of mass doing work had come into my mind, I have been pondering over: how could one think of or get the idea about this law?" general relativity asked the law of mass doing work curiously and perplexedly.

"From the famous and great $E = mc^2$, one could easily perceive or find out several fundamental questions that not only collectively and consistently, but also unavoidably and explicitly, point to the idea about the law of mass doing work. So this famous and great equation has actually made a significant and memorable contribution to the discovery of this law," the law of mass doing work replied calmly and appreciatively.

"Oh! Really?" general relativity was obviously surprised, also somewhat excited.

"Yes! Really and truly," the law of mass doing work responded calmly and assuredly.

"So I eagerly want to know why and how," general relativity reacted excitedly and hurriedly.

"You'll get what you want in this chapter," the law of mass doing work assured general relativity.

There are three fundamentally important questions, which come from $E = mc^2$, that unavoidably and inevitably, also explicitly and consistently, point to the law of mass doing work. What are these fundamentally important questions then? Why do they collectively and consistently, also explicitly and unavoidably point to this law? Let us analyze, discuss and/or investigate them in this chapter.

Question one. The famous mass-energy equivalence equation (which is $E = mc^2$, where c is the speed of light, m is the rest mass of an object, and E is the rest energy of the object) tells us that mass and energy are equivalent. Since energy can do work—being a universally accepted *fact*, why can't mass? More specifically thus to be more perceptible or noticeable, since energy has the capability of doing work—being a fully recognized *fact*, and since mass and energy are equivalent, it is thus clear, even obvious, that mass also has the capability of doing work; it is necessary and inevitable that mass has the capability of doing work. Therefore, the famous mass-energy equivalence equation (via its mass-energy equivalence and via the universally known *fact* that energy has the capability of doing work) clearly and definitely points to the law of mass doing work (because only this law can reveal, express and reflect the capability of mass doing work).

(*Commentator A: the above analysis is very sharp and penetrating! After thinking it over, I have been completely convinced that the mass-energy equivalence equation clearly points to the law of mass doing work. Moreover, this analysis is not only definitely rational and objective, but also quite simple and clear, thus pretty easy to understand and accept. And so, the above analysis could make one perceive and realize that the famous mass-energy equivalence equation has been looking forward to the law of mass doing work.)

Question two: *why* does mass have energy? The famous mass-energy equivalence equation clearly and explicitly tells us such a greatly important concept: <u>mass has energy</u>; moreover, this concept has been a fundamentally important, fully recognized *fact* in physics or in science for quite a long time. What does such a fundamentally important, fully recognized *fact* (which is <u>mass has energy</u>) clearly and explicitly tell us then? It clearly and explicitly, also unavoidably and undeniably, tells us that the question of *why* <u>mass has energy</u> is definitely and obviously a fundamentally important question in physics or in science. (Commentator B: yes, that's surely true; that's clearly true! The fundamentally important fact that mass has energy has no choice but to tell us: the question of *why* mass has energy is indeed a fundamentally important question. Moreover, no rational people in the world would deny that the question of *why* mass has energy is a fundamentally important question, as long as they are or have been familiar with the fundamentally important fact that mass has energy.)

To such a fundamentally important question, *why* does mass have energy? The answer from the law of mass doing work is: because mass has the capability of doing work. In fact, only this law can answer such a fundamentally important question, because only this law can reveal, express and reflect the capability of mass doing work. Therefore, the fundamentally important question of *why* mass has energy explicitly and unmistakably points to the law of mass doing work.

(*Commentator B: the above analysis is obviously rational and objective, also quite straightforward, thus very convincing! After deeply pondering the analysis above, I have clearly realized that the question of *why* mass has energy is really and surely a fundamentally important question in physics or in science; I have also sensed that the answer to this fundamentally important question has to depend on the law of mass doing work. As a result, the above analysis indicates or can be understood as the fundamentally important question of *why* mass has energy is eagerly longing for the law of mass doing work.)

Question three. One can think over the law of mass doing work conversely if necessary: if mass could not have the capability to do work, then it would be scientifically groundless to say that mass has energy! Or what would be the scientific basis to say that mass has energy if mass could not have the capability to do work? (Commentator C: yes, that's undoubtedly correct! If mass could not have the capability to do work, then the well-known fact that mass has energy would actually become totally groundless or baseless in truth.) And so, the well-known *fact* that mass has energy has no choice but to tell us such a clear *fact:* mass has the capability of doing work. As a result—as an explicit and noticeable result in essence, also as an undeniable or irrefutable result in truth, the clear *fact* that mass has the capability of doing work unavoidably and inevitably points to the law of mass doing work (because only this law can reveal, express and reflect the capability of mass doing work).

Moreover, if mass could not have the capability to do work, then the fully acknowledged, also widely accepted, concept of 'mass-energy equivalence' would inevitably lose its most fundamental, most essential and most important implication; this concept would thus become utterly meaningless. (Commentator C: yes, that's definitely true; that's clearly true!) Therefore, the fully acknowledged, widely accepted concept of 'mass-energy equivalence' also has no choice but to tell us such a clear *fact:* mass has the capability of doing work; this clear *fact* unavoidably and inevitably points to the law of mass doing work (because only this law can reveal, express and reflect the capability of mass doing work).

(*Commentator D: yes, that's clearly and surely true! Either the well-known fact that mass has energy or the fully recognized, widely accepted concept of 'mass-energy equivalence' has no choice but to tell us such a clear fact: mass has the capability of doing work. What's this clear fact doing now? It is eagerly looking for, anxiously waiting for, and earnestly anticipating the law of mass doing work, because only this law can reveal and show the capability of mass doing work.)

All in all, the above three questions collectively and consistently, also explicitly and undeniably, point to the objective and real existence of the law of mass doing work. And so, if this law has been discovered, such a discovery ought to be readily accepted. (Commentator E: yes, that's clearly and definitely true! In fact, what this chapter has accomplished is actually calling for the law of mass doing work; this law is also ready to come out once being called.)

# Chapter 2

# The Law of Mass Doing Work—about Why Mass Has Energy

## [The Window on This Chapter]

(Narrator: after earnestly reading chapter one, general relativity was continuing its conversation with the law of mass doing work.)
"After reading chapter one, I want to know the answers to the following four specific questions so that I can read this chapter more effectively," general relativity told the law of mass doing work directly.
"Go ahead. I'm very pleased to discuss any questions with you," this law replied kindly and gladly.
"My first question is: what's the most important task the law of mass doing work has accomplished?" asked general relativity.
"It has answered the question of *why* mass has energy by revealing and showing that mass has the capability of doing work," this law replied.
"That's great. My second question is: what's the key or decisive concept of this law?" general relativity asked further.
"Mass has the capability of doing work," this law answered concisely.
"Thank you! My third question is: what's the core principle of this law?" general relativity continued.
"It is that the amount of energy in the mass of an object is measured and determined by the amount of work done by the object's mass," this law replied concisely and clearly.
"Thank you very much! My fourth question is: what's the core point of this law?" general relativity wanted to know more.
"Its core point is: when a mass does positive work, the mass thus decreases; but when a mass does negative work, the mass thus increases," this law answered briefly.


The law of mass doing work came to the world recently, because it was discovered and verified not very long ago by me, the author of this book. (Commentator A: it doesn't matter who discovered the law of mass doing work; in fact, I don't care about the issue of who discovered this law. This is because there is such a generally acknowledged and totally accepted basic principle in science, which is also an objective and rational criterion: for any theory, the things that really matter lie in *what* rather than who—lie with *what* the theory talks about, instead of who developed it. What should be pointed out is that this basic principle has been completely recognized and admitted by the scientific community as general knowledge or common sense in science nowadays; therefore, it is definitely reasonable and realistic to believe that all today's scientists know this basic principle pretty well. On the contrary, if this basic principle were thrown away, science would inevitably lose a rational, objective and fair criterion; the most fundamental nature and spirit of science would be fatally damaged; science would definitely be misled onto a dangerous track; science would no longer be science at all! Please be reminded or please notice: when Albert Einstein found the famous and great mass-energy equivalence equation, $E = mc^2$, he was neither an important nor influential person in science, actually he was an obscure and insignificant figure in the field of science.) (Commentator B: yes! What commentator A has said above is not only definitely true, but also extremely important to the development and advancement of science, especially to the great or revolutionary breakthroughs in science.)

The core principle of the law of mass doing work is: the amount of energy in the mass of an object is measured and determined by the amount of work done by the object's mass. (Narrator or reminder: such a core principle, when viewed from the angle of comprehension, is quite comparable or very similar to that the energy of a body is measured and determined by the body's capability of doing work in classical physics. Quite obviously, also rather rationally, such a noticeable comparability or similarity is a substantial help for one to perceive and grasp this core principle easily and quickly, which can make one realize or notice that it cannot be difficult to understand the law of mass doing work, even though this law is radically new in concept.) (Commentator C: yes; what the above narrator or reminder has said is rather rational thus quite reasonable. Such a core principle indeed can readily, even easily, enable one to realize or sense: it's not difficult to understand this core principle; accordingly, it's not difficult to understand the law of mass doing work, though this law is quite new in concept.)

# The Law of Mass Doing Work—about Why Mass Has Energy

Concisely, as the exact reflection of this core principle, the law of mass doing work (with accurate mathematical expression) reveals and shows: when the velocity of an object is increased, the object's mass does positive work, the object thus loses the same amount of energy as that of the work done by the mass of the object from and by consuming its mass. As a result, the core point of this law is: an object's mass doing positive work causes a corresponding decrease in the object's mass; that is, when a mass does positive work, the mass thus decreases. (One can clearly and easily understand this core point via such a simple comparison in classical physics: when a body does positive work, the energy of the body decreases.) The other side of this core point is: an object's mass doing negative work, which occurs when the velocity of the object is decreased, causes a corresponding increase in the object's mass; that is, when a mass does negative work, the mass thus increases. (One can clearly and easily understand this side via such a simple comparison in classical physics: when a body does negative work, the energy of the body increases.) (Commentator D: such a core point is very helpful for one to perceive and realize: it cannot be difficult to comprehend this core point; it thus cannot be difficult to comprehend the law of mass doing work, though it is a newly discovered physical law.) (Commentator E: moreover, if one thinks of the core principle and the core point of the law of mass doing work simultaneously, it seems quite reasonable to believe that he or she will have no difficulty realizing it is actually pretty easy to understand this law, albeit it is a newly developed physical law.)

*Friendly reminder: dear readers, if you are the professional people in physics, especially in modern physics, you can comprehend the law of mass doing work more easily and quickly than others. This is because the law of mass doing work directly and totally comes from the relativistic kinetic energy of an object with rest mass m (by exchanging the positions of the velocity v and the relativistic momentum p in the integral calculation of the relativistic kinetic energy of the object, then by the definite integral operation from 0 to v with velocity v as variable). As a result, the relativistic kinetic energy (of the object) is the <u>area</u> that is *under* the line of a certain velocity v and *above* the line of the relativistic momentum p (i.e., the <u>area</u> between these two lines); whereas the law of mass doing work is the <u>area</u> that is *under* the line of the relativistic momentum p. That is to say, the law of mass doing work is not only connected to but also based on the known or conventional knowledge; such a feature can substantially enhance the recognition and acceptance of this law, though it is unconventionally new. Moreover, the law of mass doing work can be simply expressed as the product of the force acting on an object with rest mass m and the displacement of the object; i.e., this expression is completely consistent with the fully recognized expression of the work done by a force in classical physics. Quite obviously, also rather rationally, such a complete consistency can not only substantially but also explicitly enhance the recognition and acceptance of this law, even though it is a newly discovered physical law.

What should be pointed out is that the concept and equation of the relativistic kinetic energy of an object with rest mass m has been written into the textbooks for college or graduate education (since far more than half a century ago); that is, this concept and equation has already been fully recognized and accepted by the scientific community in physics. And so, the professional people in physics or in modern physics, because they have been familiar with this concept and equation very well, really have an obvious advantage over others in comprehending the law of mass doing work, which can make them comprehend this law much more easily and quickly than others, even though the discovery of this law might be radically new in the eyes of some of those respected conventional professional people. So my sincere congratulations go to the professional people in physics for this obvious advantage; please accept my sincere and rational congratulations if you are the professional people. (An independent and rational reviewer: because the law of mass doing work directly and entirely comes from the equation of the relativistic kinetic energy of an object with rest mass m; because this equation has been fully recognized and completely accepted by the scientific community in physics; because nothing is added or removed in the process from this equation to the law of mass doing work, it seems that there is neither rational reason nor valid basis not to accept the law of mass doing work. In fact, there is utterly no way to deny the law of mass doing work from the angle of science.)

# Chapter 3

# Revealing the Mechanism of/behind the Mass-Energy Equivalence Equation

## [The Window on This Chapter]

(Narrator: after being familiar with the law of mass doing work, general relativity wanted to know more about this law with great curiosity, thus continuing its conversation with this law.)

"Because the law of mass doing work is about *why* mass has energy; because the mass-energy equivalence equation, which is $E = mc^2$, tells us mass has energy, could this law reveal the mechanism of this equation?" general relativity asked this law with great curiosity.

"Yes, it could; because it has revealed this mechanism," this law answered clearly.

"What's this mechanism then?" general relativity was even more curious.

"It is that the rest energy of an object, being the total energy contained in the rest mass of the object, is equal to the maximum capability of the object's mass doing work," this law replied concisely.

"Then how much is this maximum capability?" general relativity went further.

"It is equal to $mc^2$, being the right side of $E = mc^2$. So this mechanism shows that the $mc^2$ (in $E = mc^2$) is not only the total energy contained in a rest mass m, but also turns out to be the maximum capability of the rest mass m doing work," this law responded specifically and concisely.

"Oh! Really? That's very interesting!" general relativity became surprised and excited.

"Moreover, this mechanism further shows that the very reason *why* the total energy contained in a rest mass m is equal to $mc^2$ is because the maximum capability of the rest mass m doing work is equal to $mc^2$," this law added more.

"Wow! That's even more interesting!" general relativity became even more surprised and excited.

As a direct application of the law of mass doing work, this law has revealed the mechanism of/behind the famous mass-energy equivalence equation (this equation is $E = mc^2$, where c is the speed of light, m is the rest mass of an object, and E is the rest energy of the object. This famous equation is often referred to as the greatest equation in the history of science).

This mechanism turns out to be: the rest energy of an object, being the total energy contained in the rest mass of the object, is equal to the maximum capability of the object's mass doing positive work (this maximum capability is equal to $mc^2$, being the right side of the famous $E = mc^2$). So this mechanism shows that the world-famous $mc^2$ (in the famous $E = mc^2$) is not only the total energy contained in a rest mass m, but also turns out to be the maximum capability of the rest mass m doing (positive) work. Moreover, this mechanism further shows that the very reason *why* the total energy contained in a rest mass m is equal to $mc^2$ is because the maximum capability of the rest mass m doing (positive) work is equal to $mc^2$.

(*Related question and answer: what is the profound and essential difference *before* and *after* revealing this mechanism? Answer: before this revealing, people merely knew mass has energy, but couldn't know *why*; after this revealing, people know *why* mass has energy via knowing that mass has the capability of doing work. Accordingly, also unavoidably, this profound and essential difference is also an explicit demonstration or clear reflection of the fundamental importance of the law of mass doing work, corresponding to the fundamentally important status of this famous and great equation in science.)

After revealing the mechanism of/behind the famous and great mass-energy equivalence equation, the solid existence of this mechanism is an irrefutable fact, a bit like: having found out the continent of North America is the irrefutable fact that there is this continent; this is also a bit like: having found out diamond beneath a certain place is the hard evidence that there is diamond beneath this place. (Commentator A: yes, no rational people in the world want to deny the solid existence of this mechanism, simply because it has been revealed; of course, no one can deny the solid existence of this mechanism, because and after this mechanism has been revealed.) Not only that, *after* revealing the mechanism of/behind this famous and great equation, its validity and reliability thus become further solid and secure—because it turns out that this famous and great equation does have a very solid and secure mechanism. (Commentator A: and so, from now on—from the moment when its mechanism has been revealed, the famous and great $E = mc^2$ can confidently and bravely declare to the whole world—with sufficient and irrefutable evidence: I do have a solid and secure mechanism!)

(Commentator B: wow! Aha! The mechanism of/behind the famous and great mass-energy equivalence equation is finally brought to light. This is definitely a remarkably important event to this great equation, because this mechanism, only this mechanism, is able to answer the biggest *why* underlying this great equation, *why* mass has energy—because mass has the capability of doing work! Moreover, if one thinks over this mechanism for a few minutes, it seems not difficult that he or she could clearly and surely realize: only the law of mass doing work is able to reveal the mechanism of/behind this great equation, believe it or not. This realization can readily, even easily, enable one to be aware that the existence of this law turns out to be an explicit *fact*, a clear *fact*, also an undeniable or irrefutable *truth*, because the existence of this famous and great equation has become a universally acknowledged, well-known *fact;* because only this law can reveal the mechanism of/behind this very equation. Thus, even if Bingcheng Zhao, the author of this book, had not discovered this law, somebody else would find it someday, sooner or later; the earlier, the better, of course. Yet regardless of who has discovered this law, the famous mass-energy equivalence equation is or should be equally happy, because this law has revealed the mechanism of/behind this famous equation, which is also often referred to as the greatest equation in the history of science. After this revealing, this famous and greatest equation, via unfolding and displaying its great mechanism that answers the biggest *why* underlying this greatest equation, also the most fundamental *why* underlying this greatest equation, appears and becomes even more beautiful and charming.)

When the famous mass-energy equivalence equation, or the greatest equation in the history of science, is at the age of more than one hundred years old, its secret veil is finally unveiled—the mechanism of/behind this famous and great equation has been at last revealed.

(Commentator C: fortunately, this famous and great equation is not a bride! Of course, if it had been a bride, probably no bridegroom in the world would have been patient enough to wait for such a long time to unveil her veil after their wedding. But for a fundamentally important and extraordinarily influential equation in science like this famous and great equation, it seems that the later unveiling its secret veil, the more wonderful its marvelous charm is. This seems to be a bit like wine: a bottle of old wine is more tasteful and mellower than a new one.) In this sense, the famous and great mass-energy equivalence equation should be happy for itself—be happy for its great mechanism having been finally revealed. In this sense, this famous and great equation ought to congratulate on itself—congratulate on its great mechanism having been at last revealed. In this sense, this famous and great equation must celebrate itself—celebrate its great mechanism having been finally brought to light! Moreover, and in a broad sense, it seems acceptable if this famous and great equation wants to invite all the people in the world, especially those respected and related experts in physics, to have a grand and solemn celebration of this great and historic revealing! (Most probably, this famous and great equation will provide delicious food and excellent wine for all of us in such a spectacular, splendid, and wonderful occasion.)

More than what we have seen above, the very fact, which is that the law of mass doing work has revealed the mechanism of/behind the famous mass-energy equivalence equation, clearly and definitely points to the great importance of this law along the following explicit and noticeable direction. Since this famous equation is widely recognized as the greatest equation in science, then its mechanism is, or ought to be, the greatest mechanism in science; since the law of mass doing work is the only physical law that reveals this greatest mechanism, then it is actually rational and appropriate (or at least it is neither irrational nor inappropriate) if one comes to the conclusion that this law is the greatest law or one of the most fundamental and most important laws in science. (Commentator D: yes; this conclusion is obviously rational and objective, thus undoubtedly appropriate. And so, it is no exaggeration to say that the discovery of the law of mass doing work is really and truly a great achievement or historic event in science, believe it or not.) (Correspondingly, what readers have seen above is, or ought to be, the greatest law or one of the most fundamental and most important laws in science; so the author of this book genuinely congratulates on you, dear readers.)

(*Related question and answer: because the law of mass doing work has revealed the mechanism of/behind the famous and great mass-energy equivalence equation, can this very equation itself be a good window, through which one could see this law more clearly, thus have a better understanding of this law? Answer: yes, it can; please see the specific and closely related facts or information in the coming paragraph.)

One could tangibly and quickly grasp the law of mass doing work if he or she views this law from the following several important, also easily perceptible, angles. Angle A, from the large perspective of the fundamental question: why does mass have energy? The answer from this law is: because mass has the capability of doing work; in fact, only this law can answer such a fundamental question. Angle B, the famous mass-energy equivalence equation tells us that mass and energy are equivalent; and since energy can do work—being a universally accepted *fact*, why can't mass? More specifically, since energy has the capability of doing work—being a fully recognized *fact*, and since mass and energy are equivalent, it is clear, even obvious, that mass also has the capability of doing work; it is necessary and inevitable that mass has the capability of doing work. (Otherwise, the fully acknowledged concept of 'mass-energy equivalence' would lose its most fundamental, most essential and most important implication; this concept would thus become meaningless in fact.) Angle C, one can think over this law conversely if necessary: if mass could not have the capability to do work, it would be scientifically groundless to say that mass has energy! In other words, the known *fact* that mass has energy has no choice but to tell us another inevitable *fact:* mass has the capability of doing work. Angle D, since mass has the capability of doing work, it becomes quite natural that, when an object's mass does *positive* work, the object's mass thereby *decreases* (one can clearly perceive and easily comprehend this point if he or she is familiar with such common knowledge in classical physics: when a body does *positive* work, the available energy of the body thus *decreases*). (Commentator E: when one views the law of mass doing work through the diverse visual angles like those above, he or she could see this law more clearly from different directions, a bit like 3-D visual effects; he/she could thus grasp this law more tangibly and effectively.)

Last but not least, what should be mentioned is that the above-mentioned fact, which is that the law of mass doing work has revealed the mechanism of the famous mass-energy equivalence equation, is also fundamentally and crucially important to this newly discovered law. This fundamental and crucial importance is explicitly reflected in such a clear *fact:* this law has been verified or confirmed via its revealing the mechanism of the famous mass-energy equivalence equation, because this famous equation has passed experimental tests many, many times since its birth, thus being a fully recognized and universally accepted *fact*. With this verification or confirmation, the validity and reliability of this law are thus quite positive. (So, for this verification or confirmation, this law, on behalf of the discoverer of this law, wants to express its sincere acknowledgment to all the related scientists for their great contributions that have made this famous and great equation found and verified, especially to the great scientist Albert Einstein, the founder of this equation.)

# Chapter 4

# The Mass Consumption

## [The Window on This Chapter]

(Narrator: having known that the law of mass doing work has revealed the mechanism of the mass-energy equivalence equation, general relativity became even more interested in this law, thus continuing its deeper discussion with this law.)

"After reading chapter two, I have noticed that the core point of the law of mass doing work is: an object's mass doing positive work causes a corresponding decrease in the object's mass; that is, when a mass does positive work, the mass thus decreases. Does such a core point have a specific, concise and accurate name or label?" general relativity was indeed very sharp and thought of a very good question.

"Your question is really very meaningful. This core point has been named as the mass consumption," the law of mass doing work replied approvingly and concisely.

"Oh, I see. No wonder the title of this chapter is *The Mass Consumption*," general relativity got the answer it wanted.

"In what condition does the mass consumption occur then?" general relativity went further.

"It occurs (to the mass of an object) when the velocity of the object is increased," this law responded concisely.

"Aha, I see. Thank you very much," general relativity replied appreciatively.

The mass consumption (being the direct result of the combination of the two things introduced in the last two chapters: the law of mass doing work and the mechanism of the famous mass-energy equivalence equation revealed with this law) shows that the mass of an object *decreases* with the increase in its velocity, by revealing *why* and *how* the object's mass is being consumed due to its doing positive work in such a situation. Concisely, the mass consumption is the *decrease* in the mass of an object, because the object's mass is consumed (becoming less) due to its doing positive work when the velocity of the object is increased. Therefore, the mass consumption, being caused by mass doing positive work, is simply the product of an application of the newly discovered and verified law of mass doing work. (Related question and answer: how could one understand the mass consumption clearly and easily, also effectively and impressively? Answer: please view and think over the mass consumption from the following three aspects or angles.)

The first aspect or angle: from the fact that mass has the capability of doing work. As analyzed, revealed and shown in the earlier chapters, mass has the capability of doing work. Since mass has the capability of doing work, it is rather natural and quite reasonable that, when an object's mass does *positive* work, the object's mass thereby *decreases* (one can clearly perceive and easily comprehend this aspect via the considerable and explicit help from such a comparable concept in classical physics: when a body does *positive* work, the available energy of the body thus *decreases*).

The second aspect or angle: from the angle of the mass-energy equivalence equation. As long as one has known or heard of the famous mass-energy equivalence equation, which is $E = mc^2$ (where c is the speed of light, m is the rest mass of an object, and E is the rest energy of the object), he or she could clearly and impressively comprehend the mass consumption, because it is inherently connected to the mechanism of this famous equation. To be further explicit, both the mass consumption and this famous equation are attached onto the same thing—the law of mass doing work, because the mass consumption, being caused by *mass doing positive work*, directly and totally comes from this law; because the mechanism of this famous equation, revealed with this law, is the maximum capability of a rest mass m *doing positive work*, as introduced in chapter three. That is, both the mass consumption and this famous equation have the same mechanism—*mass doing positive work*. As a result, this great and famous equation turns out to be a great and explicit help for one to understand the mass consumption clearly and easily, also impressively.

The third aspect or angle: from the angle of the known concept and equation. The fully recognized and completely accepted concept and equation of the relativistic kinetic energy of an object with rest mass m can also provide a substantial and explicit help for one to understand the mass consumption. This is because the law of mass doing work directly and totally comes from this concept and equation, as pointed out in chapter two; and this is because the mass consumption is simply the product of an application of this law, as explicitly mentioned in the first paragraph of this chapter.

All in all, one could understand the mass consumption clearly and easily, also effectively and impressively, if he or she views and thinks over the mass consumption from the three aspects or angles above. That is to say, with and through the substantial and explicit help from these aspects or angles, it seems rather rational or quite realistic to conclude or believe that one will have no difficulty perceiving and understanding the mass consumption (or at least have no difficulty clearly realizing the objective and real existence of the mass consumption).

Even after understanding the mass consumption clearly and impressively, some careful readers, especially some dear readers who have rich knowledge in physics, might think of the concept of relativistic mass (the so-called relativistic mass is an important concept formed within the paradigm of special relativity. This concept says that: the mass of an object increases with the increase in its velocity, and the mass of an object becomes infinitely large when the object infinitely approaches the speed of light; that is, mass increase with speed. So 'relativistic mass' is often simply said as 'rest mass is least' in the various materials on special relativity); and these readers might have noticed that the mass consumption and relativistic mass are opposite. And so, it seems better that I should provide a relevant clarification here for avoiding possible confusion. This clarification is: there are fundamental and obvious differences between the mass consumption and relativistic mass.

Concisely, these differences are reflected in the following three fundamentally important aspects. (i) The mass consumption is inherently connected with the mechanism of the famous mass-energy equivalence equation, because the theoretical basis of the mass consumption, the law of mass doing work, has also revealed this mechanism, as introduced in chapter three. On the contrary, relativistic mass has nothing to do with the mechanism of this famous and great equation, because relativistic mass was the product long before the law of mass doing work had been discovered. Moreover, relativistic mass literally prevents from revealing the mechanism of this famous and great equation. In other words, the mechanism of the famous mass-energy equivalence equation (thus this famous equation) has to say NO to relativistic mass, believe it or not. (ii) The mass consumption is totally consistent with such a fundamental principle in classical physics: when a body does *positive* work, the available energy of the body thus *decreases*. In contrast, relativistic mass is literally at odds with this fundamental principle; that is, according to relativistic mass, when a mass does *positive* work, the mass thus *increases*, which is obviously and flatly ridiculous. (iii) The mass consumption is the indispensable theoretical basis of the new theory that reveals and shows *why* time runs slower at high speed {this new theory, which is mechanism-revealed scales relativity theory, has been introduced in chapter 2 with chapter title as The New Theory Shows Why Time Runs Slower at High Speed (P. 13 ~ 30) in my earlier book with the title *Dark Matter Is No Longer in the Darkness! (The Constituents of Dark Matter Have Been Revealed)*}; whereas relativistic mass turns out to be an absolutely impassable obstacle to developing such a new theory—that is, within the paradigm of relativistic mass, it is definitely impossible for our human beings to know the secret of *why* time runs slower at high speed. All in all, with and through these fundamental and obvious differences, it seems rather rational thus quite reasonable to conclude and/or believe that one could get rid of the hindrance or interference from relativistic mass in comprehending the mass consumption.

Chapter 5

# Revealing the Mechanism behind the Postulate of the Equivalence Principle

## [The Window on This Chapter]

(Narrator: having been familiar with the mass consumption, general relativity suddenly thought of a closely related, important question to discuss with the law of mass doing work, thus continuing its conversation with this law.)

"After reading the last chapter, I have realized and noticed that the mass consumption occurs to the mass of an object when the velocity of the object is increased. And so, I was wondering whether the mass consumption has a relationship or connection with the famous postulate of the equivalence principle, considering that this postulate says that gravitational force has the same effect in increasing the velocity of an object as other traditional forces," general relativity was indeed very sharp and raised a very important and interesting question.

"Yes, the mass consumption does have a relationship or connection, actually a very close relationship or connection, with the famous, also extremely important, postulate of the equivalence principle," the law of mass doing work answered quite positively.

"What's the relationship or connection then?" general relativity became even more curious and hurried.

"This relationship or connection is, and turns out to be, that the mass consumption is also the mechanism behind the postulate of the equivalence principle," replied the law of mass doing work.

"Oh! Really? That must or should be greatly important, also remarkably interesting! And I'm exceptionally delighted that this greatly important mechanism has been finally revealed," general relativity became excited.

Being a basic postulate in the theory of general relativity, the postulate of the equivalence principle says (actually assumes) that gravitational force has the same effect in increasing the velocity of an object as other traditional forces.

# Look at the Mechanism behind the Postulate of the Equivalence Principle

What should be pointed out or realized is: to the theory of general relativity, the famous postulate of the equivalence principle is not only absolutely indispensable, but also fundamentally and crucially important—without this postulate, there would have been no general relativity at all; in fact, this postulate is literally the heart and soul of general relativity. And broadly speaking, this postulate is widely known to be one of the most important, most influential and most famous postulates in modern physics—if there had not been this postulate, the history of modern physics would have been drastically different; many parts of the history of modern physics would have been rewritten.

In the face of such a fundamentally and crucially important, absolutely indispensable postulate, it seems neither unusual nor irrational if one thinks of or asks the fundamental and crucial questions like: is there and/or what is the mechanism behind this postulate? Clearly, if there is the mechanism behind this postulate, this mechanism is (or ought to be) fundamentally and crucially important, corresponding to the fundamental and crucial status of this postulate in general relativity and in modern physics; accordingly, knowing this mechanism is not only certainly of profound and great significance, but also obviously of fundamental necessity and crucial importance. (Commentator A: yes, the above analysis is not only definitely rational and objective, but also quite simple and clear, thus pretty easy to perceive and understand.) Also clearly enough, the key to showing the existence of this mechanism lies with revealing it; that is to say, the answer to these fundamental and crucial questions hinges on whether one has revealed this mechanism.

After the mass consumption has been discovered, it turns out that the mass consumption is the mechanism behind the postulate of the equivalence principle, because the mass consumption always exists, works/performs and manifests in the situation or condition that this postulate describes or defines (which, in turn, is because the condition for the mass consumption to exist is the same as the condition that this postulate describes or defines).

Specifically, the mass consumption reveals and shows: *when the velocity of an object is increased*, the mass of the object *decreases*, because the object's mass is consumed (becoming less) due to its doing positive work; what the postulate of the equivalence principle says is that gravitational force has the same effect in *increasing the velocity of an object* as other traditional forces. Please carefully notice that the condition for the mass consumption to exist, which is 'when the velocity of an object is increased', is the same as what this postulate describes or defines, which is 'increasing the velocity of an object' (because 'when the velocity of an object is increased' is the same thing as 'increasing the velocity of an object').

Because the condition for the mass consumption to exist is the same as the condition that the postulate of the equivalence principle describes or defines, the mass consumption always exists, works/performs and manifests in the condition that this postulate describes or defines; that is, the mechanism behind this postulate is the mass consumption. Therefore, the <u>mechanism</u> behind the postulate of the equivalence principle is, and turns out to be, the mass consumption that occurs to the mass of an object when the velocity of the object is increased (i.e., when the object is accelerated), regardless of whether the increase in the object's velocity is caused by gravitational force or by other traditional forces (i.e., no matter whether the acceleration is caused by gravitational force or by other traditional forces).

(Commentator B: wow! Ah! The mechanism behind the famous and greatly important postulate of the equivalence principle is finally brought to light; this is definitely a remarkably important event to this fundamentally and crucially important postulate, because of its fundamental and crucial status in the theory of general relativity, as well as in modern physics. After revealing the mechanism behind this famous, fundamentally and crucially important postulate, the solid existence of this mechanism is a hard fact; this is a bit like: having found out the continent of North America is the hard fact that there is this continent; this is also a bit like: having found out diamond beneath a certain place is the solid evidence that there is diamond beneath this place.)

After knowing the mechanism behind the postulate of the equivalence principle, it seems rather rational and quite reasonable, at least neither irrational nor unreasonable, if one thinks of or raises the related or relevant question like: can the theory of general relativity reveal the mechanism behind this famous postulate (which is not only fundamentally

## Look at the Mechanism behind the Postulate of the Equivalence Principle

and crucially important, but also absolutely indispensable to general relativity)?

The answer to this question is explicit and positive: general relativity is unable to reveal the mechanism behind this postulate; moreover, this inability is a self-evident or actually admitted *fact*, thus also being an irrefutable or undeniable *fact*. Why? One could clearly and easily perceive and realize this inability and this *fact* via the simple and straightforward thinking like: if the mechanism behind a postulate had been revealed, the postulate would no longer have been a postulate at all; along with the specific and constant reminder from such an unavoidable *reality:* within the paradigm of general relativity, the postulate of the equivalence principle is **always** a postulate! (Related question and answer: what's the clear, even obvious, also unavoidable and inevitable, outcome of general relativity being unable to reveal the mechanism behind the postulate of the equivalence principle? Answer: clearly, even obviously, also undeniably or irrefutably, general relativity is literally incapable of reflecting this mechanism at all in all its various explanations. That is, in all the explanations of general relativity, this mechanism simply evaporates; all the explanations of general relativity are completely blind to this mechanism.)

After realizing the irrefutable *fact* that general relativity is indeed unable to reveal the mechanism behind the postulate of the equivalence principle, it seems neither unusual nor irrational if one thinks of or raises the closely related, also potentially important, specific questions like: is there and/or what is the fundamental, serious consequence of this inability of general relativity?

The answer to these questions is: yes, there is; it is impossible, definitely impossible, and absolutely impossible for general relativity to tell us *why* time runs slower and *why* length becomes shorter in a gravitational field (that is, general relativity is unable to solve the most fundamental problem in front of itself, which is *why* space and time are variable thus relative in a gravitational field), because the absolutely necessary prerequisite of being able to know *why* time runs slower and *why* length becomes shorter in a gravitational field is to find out the physical law or theory that is capable of revealing the mechanism behind the postulate of the equivalence principle. (Commentator C: after seeing the answer

Revealing the Mechanism behind the Postulate of the Equivalence Principle

above, some readers maybe feel surprised; others perhaps have the impression that your answer is too vague and general or too abstract and empty. So they may raise the question like: is there any specific, sufficient and explicit, also undeniable or irrefutable, hard evidence to show or support your answer? Answer: yes, there is; please see the specific information in the coming paragraph.)

There is indeed specific, sufficient and explicit, also undeniable or irrefutable, hard evidence showing that general relativity is really unable to tell us *why* time runs slower and *why* length becomes shorter in a gravitational field. What is this hard evidence then? This hard evidence is, and turns out to be, the plain *truth* that general relativity **absolutely** and **desperately** necessitates its indispensable postulate of 'invariant scales of length and time', which says that the scales of length and time at different points over an entire gravitational field are the same. One could easily and clearly understand this hard evidence via the simple and clear thinking like: if general relativity had been able to tell us *why* time runs slower and *why* length becomes shorter in a gravitational field, this indispensable postulate would not have been necessary at all, thus would never have appeared at all; please think over: if general relativity had been able to tell us the two *whys* above, who would have looked for trouble by proposing such a totally redundant, completely unnecessary postulate? And so, this plain *truth* is actually and exactly the hard evidence that clearly and unavoidably shows and witnesses the clear *fact* that general relativity is really unable to tell us *why* time runs slower and *why* length becomes shorter in a gravitational field.

Having known the clear *fact* that general relativity is indeed unable to tell us *why* time runs slower and *why* length becomes shorter in a gravitational field (that is, general relativity is unable to solve the most fundamental problem in front of itself, which is *why* space and time are variable thus relative in a gravitational field), it seems quite rational and reasonable, at least neither irrational nor unreasonable, that some insightful readers might think of or raise the related questions like: are there and/or what are the profound and disastrous consequences of this inability of general relativity?

The answer to these kinds of questions is explicit and positive: yes, there are; these consequences, for example, include: general relativity is not only unable to unveil the mystery of **dark matter**—this mystery includes the constituents and fundamental nature of **dark matter**, but also has actually become the shackles and obstacles to unveiling this mystery; general relativity is not only unable to unveil the mystery of **dark energy**, but also has actually become the shackles and obstacles to unveiling this mystery; because the absolutely necessary prerequisite for unveiling the mysteries of **dark matter** and **dark energy** is to find out the physical law or theory that is capable of showing us *why* time runs slower and *why* length becomes shorter in a gravitational field. As a result, the unavoidable *reality*, which is that the problems of **dark matter** and **dark energy** have actually become the two long-term unsolved, greatest problems in science within the paradigm of general relativity, turns out to be actually an impartial witness to the answer above (being the same thing as it is an impartial witness to the clear *fact* that general relativity is really unable to tell us *why* time runs slower and *why* length becomes shorter in a gravitational field).

Having been aware of the profound and disastrous consequences mentioned above, some people maybe wish that general relativity would have revealed the mechanism behind the famous and greatly important postulate of the equivalence principle; perhaps others even have the tendency to criticize general relativity for having not revealed the mechanism behind this postulate. However, I want to point out or clarify that general relativity shouldn't be criticized for having not revealed the mechanism behind this postulate; because the reason of having not revealed the mechanism behind this postulate in the past was that the law of mass doing work has been discovered too late. Please carefully notice that the mass consumption, which has been proven to be the mechanism behind this postulate, is simply the product of an application of the newly discovered and verified law of mass doing work, as clearly mentioned in the first paragraph of the last chapter; and please also carefully notice that this law came to the world about one century later than general relativity. Therefore, one should not criticize general relativity for having not revealed the mechanism behind this postulate.

# ABOUT THE AUTHOR

Bingcheng Zhao, who was born in 1963 in Shandong Province of China, obtained his Ph.D. in 2001 from Washington State University. He is the author of the popular science books: *Why It's Difficult to Understand "A Brief History of Time"*, published in 2016; and *Dark Matter Is No Longer in the Darkness! (The Constituents of Dark Matter Have Been Revealed)*, published in 2018. He is also the author of the academic book: *From Postulate-Based Modern Physics to Mechanism-Revealed Physics*, published in 2009; the newly developed and verified mechanism-revealed physics is the key to solving the fundamentally important problems of dark matter and dark energy; and the birth of mechanism-revealed physics actually heralds that the spring of science is coming again.

www.ingramcontent.com/pod-product-compliance
Lightning Source LLC
Chambersburg PA
CBHW031531210526
45463CB00010B/3052